SCIENCE PLEXERS®

A Collection of WORD PUZZLES

Biology, Chemistry, Geology, Meteorology, Physics, Pseudoscience

David Hammond

Tom Lester

Joe Scales

DALE SEYMOUR PUBLICATIONS®

This book is published by Dale Seymour Publications®,
an imprint of Addison Wesley Longman, Inc.

Dale Seymour Publications
299 Jefferson Road
Parsippany, New Jersey 07054
Customer Service: 800-872-1100

Managing Editor: Catherine Anderson
Production/Manufacturing Director: Janet Yearian
Sr. Production/Manufacturing Coordinator: Fiona Santoianni
Design Director: Phyllis Aycock
Design Manager: Jeff Kelly
Text and Cover Design: Nita Ybarra Design
Composition and Computer Graphics: Eileen Sullivan

Order number 27305
ISBN 0-7690-0128-9

1 2 3 4 5 6 7 8 9 10-ML-03 02 01 00 99

DALE SEYMOUR PUBLICATIONS®

This Book Is Printed On Recycled Paper

ACKNOWLEDGEMENTS

★

We would like to thank the Sacramento Bee for eighteen years of support by having a Plexer appear daily on the comic page. We appreciate the thousands of teachers and students who have made Plexers a part of their school programs, both as an academic tool and as a mind-expanding recreational pastime. We have published nearly 6,000 different Plexers. The encouragement to produce topical books came from subject matter teachers who through the years have asked, "Don't you have Plexers that I can use with my students?" You are holding the result. We trust that we have met your expectations. To all of you Plexer fans, a big thank you!

David Hammond
Tom Lester
Joe Scales

INTRODUCTION

★

PLEXER is derived from the word *perplex* which means; "to confuse or puzzle; bewilder, to make confusedly intricate." It is the the prime objective of Plexer books to encourage students to think logically and to expand the ways in which they view the world and solve problems.

Divergent thinking is at the heart of Plexers. It gives students another communication code. Plexers allow them to try different solutions without risk, since there are no grades when solving Plexers. The possibility of developing an original piece of work is very high with Plexers. To invent something no one has ever seen before is very exciting.

Plexers are always associated with a Plexer box. The words and symbols placed in and around it stand for expressions, idioms, people, events, objects, and, in general, things that students are familiar with to some extent. The Plexer box gives license to use spelling and grammatical errors that are not acceptable in students' daily work. This has an appeal that motivates even the most precise student.

It is important that those who know the solution to a Plexer do not spoil the enjoyment of others discovering it on their own. An essential stage of problem solving, the "incubation stage," is rare in America's classrooms. Teachers are so used to "telling" and having closure that they forget students need to develop internal processes that advance data into a sequenced understanding. It is important to instruct for process as well as knowledge.

Using Science Plexers in the Classroom.

- Science is filled with symbols, persons, and processes that lend themselves to Plexers. After students have had some experience solving the Plexers in this book they can create their own Plexers using topics they are currently studying or in which they have a special interest.
- Teachers may post a group of 2 or 3 Plexers on the bulletin board. At a convenient time of the day, ask students if they have a conjecture of what any of the Plexers mean. Difficult Plexers may stay on the board for a couple of days or longer. Invite students who have solved one or more of the Plexers to provide a new hint each day for those who have not.
- Include Plexers on worksheets, tests or quizzes as a last question or as a problem solving item. Some Plexers may have several solutions. Make a list of them and discuss the merits of each solution. There are no right or wrong answers. If the solution meets the Plexer's symbolism, it is acceptable.

- Provide a page of Plexers to students. Have them work individually, in pairs, or in groups to solve the Plexers. Have students explain how they solved each Plexer. Use the discussion to refine problem solving strategies.
- Have students identify types of Plexers, giving samples of each type. For example, "location" could be one of the types. Under "location" students would include examples of the location of the Plexers in relation to the Plexer box: on the sides, down low, up high, in or out, in the middle, or even in corners. Students soon realize that location often becomes part of the solution for many of the Plexers.
- Use Plexers at the end of a class to keep minds occupied and promote mental development. Plexers also make good warm-up activities when returning to school after a long holiday.
- Have students choose their favorite Plexers and tell why they are favorites. This provides the students an opportunity to express their personal preferences for a variety of thought processes. Students will often choose one they created as their favorite, reinforcing the importance of being creative. Plexer creation is a form of creativity such as creative art, music and writing.
- As a general rule, the difficulty level of the Plexers increases from the beginning to the end of a page.
- Most importantly, Plexers are for fun! It is the authors' fondest wish that everyone enjoy the challenge of Plexers.

1

Y
L
L
E
B

2

MOON

3

shkeepape

4

MOTION –

5

H E
O L

6

" NATURE

7

CHOVY

8

DigeHClstion

9

I
IC

10

♥ BURN

11

CHO–TEROL

12

A GE

13

14

B
A I
R N

15

R A I N

R A I N

16

DAWN

17

MON C TER
MON C TER
MON C TER

18

TRUM

19

P
o
u
r

20

PUT
lbs

21

NEPAINCK

22

ATOM

23

APP E DIX
N

24

PE RS
ALITY

25

C † CE

26

+hough+
+hough+

27

as gus
gus

28

R NIAP

29

DK

30

RU
WEIGHT

31

DAY

32

+ ER X

33

M I C R O

34

π CC

35

CE CE
CE CE

36

L O D A

37

site
site

38

$ky $ky
¢a$t

39

DEAD
ARRIVAL

40

SCIE NCE

41

B9 mor
mor

42

TUMDRN

43

O n

44

SWALLDOWN

45

¢ORY
CEPTION

46

TAT
TAT

47

TIDE

TIDE

48

VI P ER

49

FEC POX TION

50

10TACLES

51

PMUH
WHALE

52

8
VAC
VAC

53

R R
A
Y Y

54

QTZ

55

RAT

56

FEET

57

2th DK

58

CING
★ ★ ★

59

RECORD°

60

EGAMI

61

BOT BOT BOT

62

MD
CALL

63

D
♂
O
R
S

64

Me † der

65

H T
I I
B
I I
H T

66

DEAD
———
2

67

T
H
R
O
MEDICAL
U
G
H

68

H E
R
B

69

UNIVERSE

70

SEA¢ON
¢EASON

71

¢ part part ¢

72

VISION
V V
I I
S S
I I
O O
N N
VISION

73

DEEFOIB

74

LAX
LAX

75

TAX
OMY

76

MAT MAT

MAT MAT

77

H_2O AGE

78

T
E
A
R

79

80

PE
PE
PE

81

GALA CTIC

82

B4 S U N

83

$UU

84

CEP
CEP
CEP

85

JE T

86

UR
?
U
EAT

87

♂ GO

88

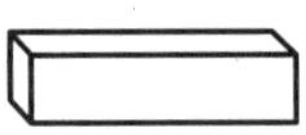 LEE

89

S5280'S

90

T H E R I A

91

♥ED

92

Ƨ

NATURE

93

F

FIRe

e

94

☒ PAY DIRT

95

+ ICTION

96

S¢

97

e t

98

W I N G

99

♂ C

I

100

T T
I C
N
I C
T T

101

EAR
EAR WET

102

WIRE —
RADIO

103

SLPOASCTE

104

W
I
WI RE
R
E

105

GRA ITY
V

106

Fe E

107

PER10SION

108

HUO
BING

109

BUR

110

VOLTAGE

111

MEAL BRUSH
MEAL BRUSH
MEAL BRUSH

112

Re¢ x

113

W♡♡D

114

BSIECKD

115

t A g

116

dr8

117

PERSONALITY

118

10
DER
12″

119

TOOTH T

120

+ ER

121

SP wt.– ACE

SCIENCE ★ PLEXERS®

122

SHOWER
SHOWER
SHOWER
SHOWER

123

~~tion~~

124

E ¢ TRICK

125

MY
GNIHCA

126

NOISE –

127

HE
8

128

graduate°

129

★ VATION

130

CCCC

131

$WHEAT

132

$MERE

133

NOIA
NOIA

134

= ii

135

CORN
COB

136

A L P H A

137

Au 2th

138

I
OSPHERE

139

Ba $(Na)_2$

140

♂ed
CRAFT

141

Ag FOX

142

R
O
C
K

R
O
C
K

143

ATOM
ATOM ATOM ATOM
ATOM ATOM ATOM
ATOM ATOM ATOM
ATOM ATOM ATOM
ATOM ATOM ATOM

144

NIG

H

T

145

HO

Ag

146

FOOD

+ ITIVE

147

G
A
R
B
A
G
E

G
A
R
B
A
G
E

148

DISK

149

PLA¢TA

150

AIR

151

MO♂TUM

152

WRC_2H_5OHIC

153

K9

154

W
I
N
D
E
R

155

L NK
LI K
LIN

156

COAST

157

¢ORY

158

P P
I I
R R
E E

159

LENS
LENS

160

†AC
DA

161

I ARS C

162

i i
O O

163

SIGHTED

164

NIAP

165

OAT

SCIENCE ★ PLEXERS®

166

E O Z O I C

167

A K I T

168

O
X

169

ME ME
C C
UM UM

170

VO

CANE

171

POODLE

172

Wo♂

173

Fe π
RITE

SCIENCE
PLEXERS®

174

L
E
E

175

BITE

176

@ om
@ om

177

DRILLING
SHORE

178

SNO

W

179

R O W O

M R

R M

O W

W O

M R

R M

O W

180

~~BLE~~

181

W

WO

WOLF

WOLF

WOLF

LF

182

L L

E E

G G

183

EDIFOODBLE

184

CEP CEP

CEP CEP

185

HAERT

186

GRADE°

187

188

SLEP

189

FIN
FIN

190

P I T
MINE

191

ETATS
STETA
STAET

192

SONSON

193

SHORE
CURRENT

194

195

196

197

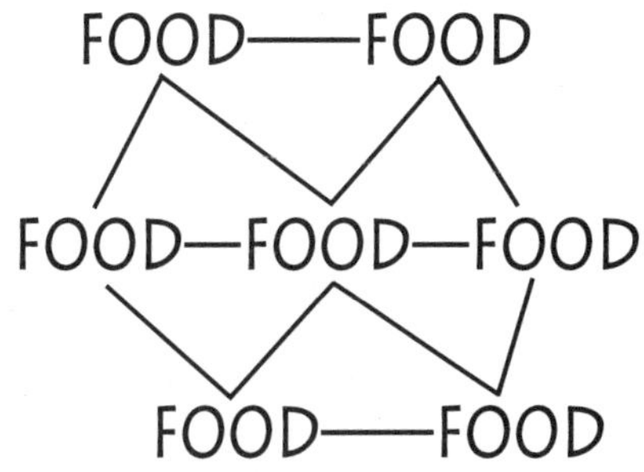

198

TIFSCIIC

199

Pb'

200

ENTENT

201

A
T
U
R
E

202

?NAIRE

203

DANGSPECCERED

204

PLASnUM

205

EAGLE

206

ER

207

DRINK

MANY
MANY

208

E E

Y

E E

209

VERENT

210

HA IR

211

DAYDAY

212

F A C E

213

DCC

214

WEIGHT

215

CRE 8

216

C
O
M
EARTH
EARTH

217

CONSnENTAL

218

♂ tis

219

ORCHARD
ORCHARD

220

WETHER

221

An
E
GO

222

IC LE

223

SHOWER
SHOWER

224

RIVER

225

FEVER

226

CCC
AL
LERGG

227

FEED
DOD

228

P
O
ATPOLLION
L
L

229

NATALIION

230

CLAgOUAgD

231

+ CRE8

232

S & $

233

SABER
2th ed
CAT

234

DO ·

235

ROAD

236

AIR

237

SYTHIA
SYTHIA
SYTHIA
SYTHIA

238

OF SYNC

239

Phen
enon

240

DNUORG
RADIATION

241

WI DE

242

LEG

243

STEP
4
♂

244

CABIN°

245

B
B Y
BY-PASS
P P
A A
S S
S S

246

DRUG
1,2,3...er

247

PUT
lb lb lb

248

✓
HURRICANE

249

H
C
T
A
C
SLEEP

250

MOON

251

WAT ER

252

ROZAR
HOG

253

B
YOUR
E
A
MOUTH
H
E

254

Au

255

ETAIGN

256

D

DR8

257

S
T
WINOTER
R
M

258

LYSE
LYSE

259

MIND
MATTER

260

2th –

261

WEATHER
BB

262

DROUGHT

SCIENCE PLEXERS

263

S
N
MEALAMEAL
C
K

264

♂ URE

265

VENTREING

266

SIR

267

H_2O $\$$

268

PER4♂GE

269

EUR LOUIS

270

8 (G E R M)

271

GOT
BB
GOOD
UUUU

272

WEA¢THER

273

SCRAP SCRAP
SCRAP SCRAP
SCRAP SCRAP SCRAP
SCRAP SCRAP SCRAP
SCRAP SCRAP SCRAP
SCRAP SCRAP SCRAP
SCRAP SCRAP SCRAP

274

¢ ORGAN

275

BOR8

276

SEC
OND

277

H
2
O

278

♂ tle

279

EST EST

EST EST

280

CAN

CAN

281

LACND

282

HELIX

283

R
E B
E
Z E

284

SPEED

ANSWERS

★

page 1

1. belly up
2. full moon
3. keep in shape
4. motionless
5. sink hole
6. second nature
7. anchovy
8. acid indigestion
9. ionic

page 2

10. heart burn
11. high cholesterol
12. space age (the)
13. half a notion
14. scatter brain
15. rain drops
16. the crack of dawn
17. sea monsters
18. spectrum

page 3

19. downpour
20. put on weight
21. a pain in the neck
22. split atom
23. ruptured appendix
24. split personality
25. seance
26. positive thoughts
27. asparagus

page 4

28. lower back pain
29. decay
30. are you overweight?
31. daybreak
32. summer times
33. microwave
34. Pisces
35. force
36. elodea (aquarium plant)

page 5

37. parasite
38. sky's overcast
39. dead on arrival
40. space science
41. benign tumor
42. medical intern
43. lion
44. down in one swallow
45. sensory perception

page 6

46. tattoo
47. high and low tides
48. pit viper (snake)
49. smallpox infection
50. tentacles
51. humpback whale
52. vaccinate
53. X-ray
54. quartz

page 7

55. cornered rat
56. flat feet
57. tooth decay
58. seeing stars
59. record high temperature
60. reverse image

61. robots
62. doctor on call
63. outdoorsmen or outdoorsman

page 8

64. meander
65. exhibit
66. half dead
67. medical breakthrough
68. herb tea
69. expanding universe
70. change in the seasons
71. interchangeable parts
72. peripheral vision

page 9

73. biofeedback
74. parallax
75. taxonomy
76. format
77. water shortage
78. teardrop
79. nickel
80. tripe
81. intergalactic space

page 10

82. before sunrise
83. cashews
84. triceps
85. jet lag
86. you are what you eat
87. mango
88. barley
89. smiles
90. diphtheria

page 11

91. downhearted
92. back to nature
93. irons in the fire
94. strike pay dirt
95. addiction
96. essence
97. eternity
98. wing spread
99. manic depression

page 12

100. extinct
101. wet behind the ears
102. wireless radio
103. lost in space
104. crossed wires
105. low center of gravity
106. irony
107. hypertension
108. human being

page 13

109. back rub
110. high voltage
111. brush after every meal
112. recent times
113. heartwood
114. sick in bed
115. anagram
116. hydrate
117. warped personality

page 14

118. tenderfoot
119. chipped tooth
120. adder (snake)
121. weightless in space
122. scattered showers
123. notion
124. eccentric
125. my aching back

page 15

126. noiseless
127. he over ate
128. graduate degree
129. starvation
130. foresee
131. buckwheat
132. cashmere
133. paranoia
134. equalize

page 16

135. corn on the cob
136. alpha wave
137. gold tooth
138. ionosphere
139. banana
140. manned spacecraft
141. silver fox
142. fallen rocks
143. atomic pile

page 17

144. night fall
145. hi ho silver
146. food additive
147. garbage in, garbage out
148. slipped disk
149. placenta
150. air tight
151. momentum
152. "workaholic" (it could be!)

page 18

153. canine
154. sidewinder (snake)
155. missing links
156. the coast is clear
157. sensory
158. expire
159. concave and convex lenses
160. anaconda (snake)
161. arsenic

page 19

162. dark circles under the eyes
163. near-sighted
164. back pain
165. manatee
166. archeozoic
167. tourniquet (turn a kit)
168. onyx ring (a ring on X)
169. paramecium

page 20

170. novocaine
171. miniature poodle
172. woman
173. iron pyrite (fool's gold)
174. portly (left lee)
175. big overbite
176. atoms
177. offshore drilling

page 21

178. snow fall
179. ringworm
180. Noble, Alfred B.
181. wolf pack
182. bowed legs
183. inedible food
184. forceps
185. heart disorder
186. low-grade temperature
or low-grade fever

page 22

187. nova
188. inadequate sleep
189. paraffin
190. open pit mine

191. altered states
192. bison
193. longshore current
194. Lady bug
195. open-heart surgery

page 23

196. hands-on science
197. the food chain
198. scientific
199. lead foot
200. apparent
201. curvature
202. questionnaire
203. endangered species
204. platinum

page 24

205. bald eagle
206. the big dipper
207. one drink too many
208. cross-eyed
209. reinvent
210. parted hair
211. the days are getting longer
212. face lift
213. disease

page 25

214. overweight
215. create
216. come down to earth
217. continental divide
218. mantis
219. pear orchard
220. a poor spell of weather
221. an eon ago
222. particle

page 26

223. diminishing showers
224. the river is rising
225. the fever broke
226. seasonal allergies
227. feed on demand
228. cross pollination
229. alien nation
230. cloud with a silver lining
231. procreate

page 27

232. sand dollar
233. saber-toothed cat
234. dew point
235. road-cut
236. compressed air
237. forsythia
238. out of synchronization
239. phenomenon
240. background radiation

page 28

241. open wide (dentist request)
242. broken leg
243. "One small step for man..."
244. cabin fever
245. triple by-pass
246. over the counter drug
247. put on a few pounds
248. check the eye of the hurricane
249. catch up on sleep

page 29

250. the man in the moon
251. water gap
252. razorback hog
253. breathe through your mouth
254. heart of gold
255. eating disorder
256. dehydrate
257. mid-winter storm
258. paralyze

page 30

259. mind over matter
260. toothless
261. to be under the weather
262. the drought is over
263. a snack between meals
264. manure
265. reinventing
266. circuit

page 31

267. waterbuck
268. performance
269. Louis Pasteur
270. germinate
271. got to be good for you
272. a change in the weather
273. scrape heap
274. sense organ
275. borate

page 32

276. a split second
277. water right
278. mantle
279. forest
280. toucan
281. inland sea
282. double helix
283. sea breeze
284. warp speed